DIAMOND PROSPECTING IN VIRGINIA AND WEST VIRGINIA

Source of the Punch Jones Diamond Found and Theory of Diamond Formation

By

Ronald N. Bone, Ph. D.

Copyright © 2013 **Ronald N. Bone**

All rights reserved.

ISBN: 1493689673
ISBN-13: 978-1493689675

Ronald N. Bone

DEDICATION

To all who have supported Mi Vida Loca.

Ronald N. Bone

Diamond Prospecting

Ronald N. Bone

Table of Contents

BACKGROUND AND PERSONAL NOTES 1

PROSPECTING 13

GEOLOGICAL HISTORY 26

PROSPECTING ON THE CHEAP 33

DIRECTIONS TO THE PLACES COVERED ABOVE .. 39

TECTONIC THEORY OF DIAMOND FORMATION 47

CONCLUSION 51

APPENDIX A 1

   CONTACT INFORMATION 1

APPENDIX B 2

   BIBLIOGRAPHY 2

APPENDIX C 4

  Glossary of Scientific Terms 4

About The Author 9

Ronald N. Bone

Acknowledgements

INVESTORS

Dr. Lloyd W. Cowling

Dr. Alan C. Lazer

J. B. Buckland

George Gillespie, O. D.

Charlene Lewis

Betty R. Rader

PROSPECTORS

Jeanne R. Bone

Jerry "Topper" Averill

Tommy Thacker

Delmis Keen

Mark Asper

Kevin Averill

Sandy Averill

Photographs courtesy of Sandra Sears Cowling, with the exception of the Jones Diamond from Sotheby's Inc.

Ronald N. Bone

BACKGROUND AND PERSONAL NOTES

This book is intended to provide the reader with information regarding diamond prospecting. It also points out areas to prospect. I have prospected in this area of Virginia and West Virginia in the past, but my ill health prevents me doing it anymore. It's all yours now. Finding diamonds will not be that difficult, but remember, they should be two carats or larger and free of flaws to be worth much. You may want to save everything smaller, too. Some third world company may purchase them from you.

The big money lies in colored diamonds which require a lot of time inspecting small stones. The market is good for colored stones. Be on the lookout for stones coated in red iron ore from a layer on top of Tuscarora Sandstone (TS).

Diamond prospecting is legal in the Jefferson National Forest. But, you can't bring in big equipment for serious mining claims there.

One thing you will have to put up with eventually, if prospecting becomes a popular hobby, is finding those secret sites that prospectors will take pains to hide. There will be no central agency to report sites which could be false. If you go diamond prospecting, be aware of possible thieves in the area. You should check in with the authorities when you enter or exit remote areas as a safety measure in case you become injured or lost. Forest rangers can help.

If you are interested in learning more about diamonds, these two older books are excellent references:

<u>Diamonds from Birth to Eternity</u> by Arthur N. Wilson (1970), and <u>The Nature of Diamonds</u> by George E. Harlow (1998).

My interest in diamonds started when I was in the fifth grade. I was at my grandmother's house, which had a good selection of books. I came across a pamphlet on West Virginia minerals. To my surprise, it included a diamond found in Peterstown, WV.

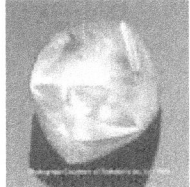

Photograph courtesy of Sotheby's Inc. (c) 1984.

The diamond weighed slightly over 34 carats and probably was the largest diamond every found in the United States. A bunch of boys were playing horseshoes when a boy named "Punch" Jones discovered it. He was from a large family and kept the stone with his treasures and forgot about it.

Years later, Punch attended college at Virginia Polytechnic Institute (VPI or now, Virginia Tech) in Blacksburg, VA. He took his stone to the geology department where it was identified as a diamond.

No one knew where it came from or why it was in a small town in West Virginia. Some speculated that it was from an Ice Age glacier deposit, others guessed it was dropped by an Indian long ago. I was surprised but thought little of it until 1985.

Up until this time, I was more interested in gold, mainly by reading about gold mining the West. I had become less interested in gold after I learned that they used both cyanide and mercury—two deadly poisons, to extract gold from ore. Panning for gold was a better way to prospect since it disturbed little in the environment and didn't involve polluting with toxic chemicals to extract it.

During the Depression, I would imagine that almost every stream in West Virginia and elsewhere were prospected for gold. Any diamonds found were either discarded or saved as trinkets. More kids than you think, could have found and kept diamonds while playing in stream beds. No one will ever know how many diamonds were found while panning for gold. If you check county records, a number were found in North Carolina in different locations.

It is amazing how things are overlooked. One man in North Carolina kept a large chunk of gold as a doorstop. Another man in North Carolina found an emerald and was fortunate enough to sell it and purchase a small island he could live on.

This country has single-mindedly focused on gold. Conversely, diamond prospecting could become a craze if people realized how profitable it could be. In 1985, I was with my friend Maurice Griffin in Myrtle Beach, SC. He was recovering from an operation while I was checking out the local library for geology books and read a few on diamonds. Of course, the story about the Punch Jones Diamond caught my interest because I come from Bluefield, WV—only thirty miles from Peterstown, WV, where the diamond was found. We were heading back to Buckhannon, WV, so I continued to read everything I could find on diamonds from the libraries there, both public and at West Virginia Wesleyan College. While I was visiting my mother in Bluefield, I researched diamonds some more with inter-library loans in two states. In addition, I mailed post cards to scientists publishing on diamonds, requesting copies of their research.

To summarize what I learned about diamond formation, Kimberlite pipes appear to emerge through cratons that lie 90 to 120 miles below the surface of the earth. As erosion progresses, the kimberlite is reduced and deposit its diamonds in sandstone layers. This is called a secondary deposit. Specifically, the sandstone deposit I was fascinated with is called Tuscarora Sandstone (TS) and it covers both East River Mountain and Peters Mountain. They are actually the same mountain, bisected by the New River on the West Virginia – Virginia border (Glen Lyn, VA) not far from Peterstown. Both of these mountains contain diamonds.

East River Mountain to the left and Peters Mountain to the right

Shot from Narrows, VA looking toward Glen Lyn

Nearby, in Tazewell County, VA, other diamonds were found by a farmer in the early 1900s. A flash of light on the ground caught his eye while he was plowing his fields. The diamond he found was called the Gillespie Diamond. It was cut and polished and placed in a ring.

Another diamond was found near Richmond, VA in the 1800s by some men digging a ditch. It was found deposited in the James River. It probably was carried there by Potts Creek. Potts Creek flows from Peters Mountain.

A trip to VPI in Blacksburg, VA revealed a kimberlite pipe, which was later found to be a lamporite pipe which is slightly different chemically. Lamporite pipes have a wide mouth where as Kimberlite pipes have a distinctive shape like a carrot.

The pipe was found in Rockbridge County, VA. This county is named for the Natural Bridge and contains the towns of Lexington and Buena Vista, as well as Natural Bridge. Only the bottom was found, and no diamonds were recovered. Included Pyrope Garnets are known as a source rock which always occurs with diamonds. These were found in the Lamporite pipe, but no diamonds with them. Several other minerals also occur with diamonds, so this is probably the source of diamonds in TS.

The Lamporite pipe was dated at 440,000 years ago (the same age as the TS). Don't take these dates as exact because these dates come from a normal bell curve of the most probable age. It could be off by several years. Recently, better dating methods have been developed. Many geological time periods have been re-determined. It is, however, as close as we can guess for the Lamporite pipe and the TS at this time.

A shining example of the Lamporite pipe found in Western Australia has produced a huge volume of diamonds. There were so many that they were shipped to Bombay, India to be cut and polished by

800,000 Indians who put facets on the stones with relatively primitive means.

Diamonds are graded by their size and quality. The pros use the 4 C's when classifying and pricing diamonds: carats, color, cut, and clarity. Nine tenths of all stones are used industrially for cutting and grinding equipment. Leaving 1/10 which are gem quality stones used in jewelry. Australia bragged that they find twice the quantity of gem diamonds of anywhere else in the world—or 1 out of every 5 stones are gemstone quality. As a consequence, these diamonds account for a large portion of the gemstones in circulation today but many are cheap in value.

Lamporite pipes are noted for industrial diamonds, but also some very nice gems with good color (higher value) are found in them. Since the colored stones hit a big price in the late 1990s, prices have dropped considerably, but are still worth far more than clear diamonds.

Diamonds form at great depths below the earth's surface, under great pressures and heat. Geologists estimate that most diamonds form in a range of 90 to 120 miles deep. Cubic diamonds are the first to form nearer the surface. These start taking on more facets up to eight-sided gems in the middle layers, and go up to twelve-sided gems at the most extreme depths.

Strange things emerged from my research. Many-sided diamonds can form cubes sprouting from the main stone. Colored diamonds have different mineral impurities that give them their distinctive over-all color. Blue diamonds contain lots of boron, while yellow ones have dominant nitrogen molecules scattered inside them. There are also red and orange tinted diamonds. All diamonds conduct heat and are the hardest substance known to man. This is why industrial diamonds are used as cutting materials for drills and augers. Two books that I mentioned in the Preface have good reviews on this subject, and proved invaluable on my many diamond prospecting trips. Knowledgeable research on diamonds is published in a number of geological journals. Chief among these that I prefer is called Science and Nature. These can be found online along with many other lesser known journals.

I took some time off and travelled out West and pretty much "lived" in libraries. After a period of time, I compiled a long list of researchers and requested copies of their articles. On a visit to see my friend Maurice Griffin in Merida, Yucatan in Mexico, I sent off hundreds of requests all over the world for more diamond research (mostly to Japan, Russia, and Australia).

Mexico is famed as an easy-going country, which seems to apply to its postal system as well. It felt like a whole month before their mailed replies reached me there. The complex where we lived was huge, surrounded by a wall, three swimming pools and a deserted old house of ill repute. My friend rented his place for $25 a month. It was an empty place with marble features and even a bidet. I really enjoyed reading all the research articles there.

An important thing occurred to me while Maurice and I were discussing things and looking at a map of the Atlantic Ocean (minus the water). I noticed that the floor of the ocean had transverse faults running both east and west of the Mid-Atlantic Ridge of a mountain. A funny feeling overcame me and I left the room to lie down to meditate and think of nothing (not easy for me).

In a little while, my mind turned to the ocean floor and it dawned on me this was connected to diamonds in some manner. A number of years ago researchers learned that the carbon in diamonds is mostly carbon-14. There are also small amounts of carbon-13 which is derived from living organisms. How did it get there, if not from subduction? This could have happened when the earth's tectonic plates pushed together, and one forces the other under it. Later I found this insight to be significant in my theory of diamond formation years later.

After the Yucatan, I traveled by bus to San Diego, CA where I stayed in a motel near the university there. I headed to the card catalog in their library only to find they had switched over to a computer system catalog. Time can slip up on one, and you find yourself lost in technology.

I ran across an article about diamond prospecting in British Columbia, Canada. A geologist named

Fipke, who circa 1991, examined a trail of Pyrope garnets left from the advancing ice sheets during the Ice Age. These ice flows scraped the top off a Kimberlite pipe. It took a lot of time to trace it to its source which was far north in the Canadian Northern Territories. Today, it is a great producer of fine quality diamonds.

I bought a truck in New Mexico and spent about six months camped on a ledge overlooking a deep canyon. Then I visited a friend in Marble, CO named Chris Barr. An interesting thing occurred when I met a real mountain man who lived alone but came out every now and then to just get drunk and have a good time. He mentioned a place called Dotsero, CO, where there are diamonds, but they were just black marbles and not worth a damn! My truck was very difficult to start in cold weather, and I never made it to Dotsero. Winter time and late autumn are not a good time to be in Colorado, so I returned to West Virginia.

Ronald N. Bone

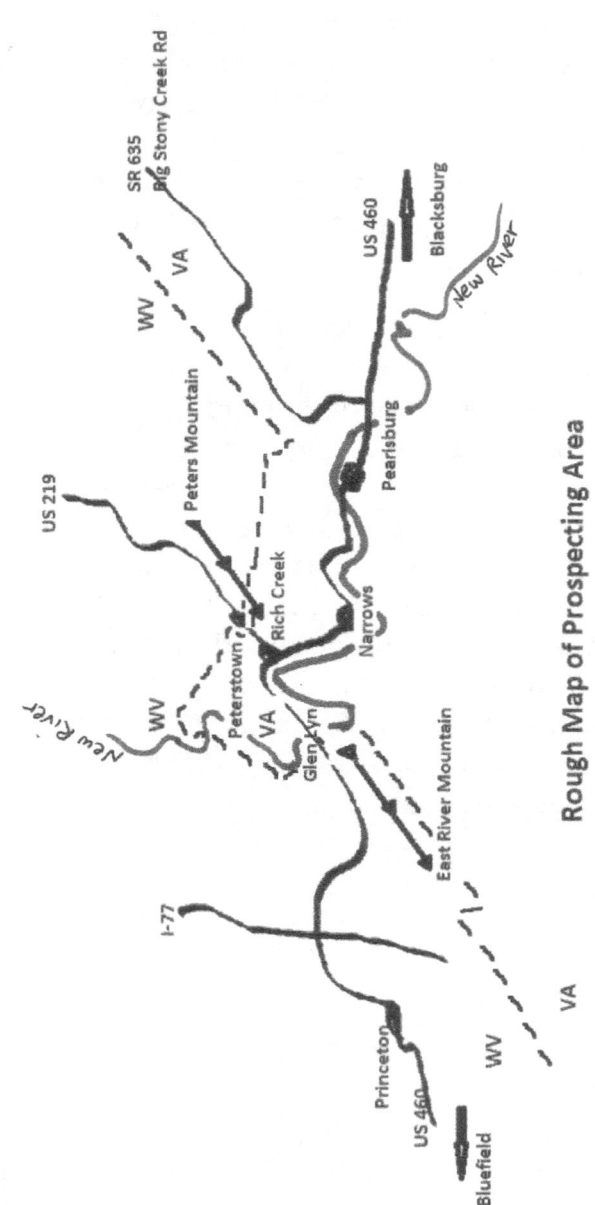

PROSPECTING

Diamond Prospecting

To make a long story shorter, my wife and I got back together and we moved to Peterstown, WV. Since we needed to find a place to rent, we drove all over town to talk to folks. One of the best ways to find out about older times is to talk to older people. I found them more than willing to talk to someone new. One story really intrigued me. It seemed two prospectors who were probably working Rich Creek, which runs through Peterstown, at the time, found several diamonds. These stones were not of high quality, so they gave up. Rich Creek (the creek, not town) was where the Punch Jones Diamond was found.

We finally found a place to stay and rented from

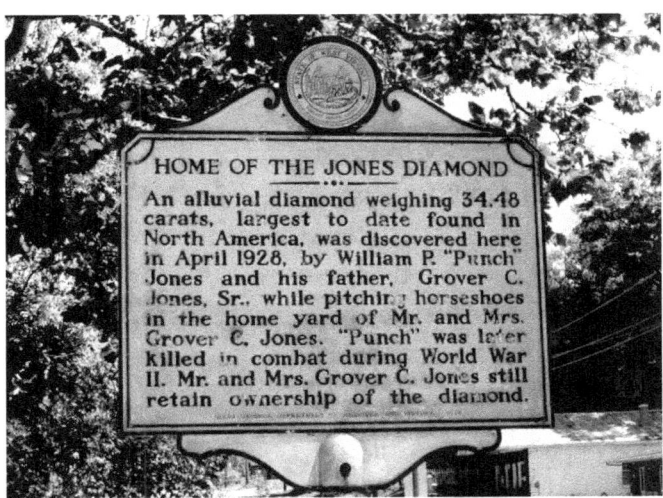

Tad Jones, who was a brother to Punch Jones. My wife had a distant relationship with a lawyer in Peterstown, and I visited and had an interesting talk with him. He told me about a geologist who came to Peterstown to search for diamonds. He finally brought it to a stop and said all he found were the industrial grade stones. Remember, 9/10 of all diamonds found are used in industry. So, there HAD been a follow up to the Punch Jones Diamond find.

We rented a house where Peters Mountain could be viewed from the back door. It doesn't look like it is that high, but try climbing it! I did, and I thought I was going to die. In reality, it is HUGE and climbs roughly 3,800 feet and over 4,000 feet in places.

The west side of the mountain is owned by farmers. The top of the mountain forms the state boundary with Virginia, and contains the Jefferson National Forest. Prospecting is legal in this area, but you cannot use machinery, or bring in a mining company.

The eastern side of Peters Mountain has a gentler slope to it than the western side. Some streams drain off the eastern side and would provide a nice quiet place to prospect. During warm periods, a lot of people walk the ridge on the Appalachian Trail. The streams' water ends up in Stony Creek which flows south to the New River.

The first time I prospected there, I had my wife drop me off on a dirt road that ran over Clendennin Creek. It was a long walk, but I finally came across a nice spot where sediments were piled up. Being an amateur, I used several strainers with the fine holes the size of a screen door mesh. I had two half gallon motor oil cans which I filled with my finds and carried out. At home, I took three or four sheets of newspaper and laid them out in the back yard to spread the sediments evenly over the paper. It didn't take it long to dry. Then I placed the resulting sand in a bucket where it was easy to work with. I ended up with so much waste material, that I dumped it all in the driveway. It could be the only driveway in the US that contains industrial diamonds. Incidentally, they are difficult to spot, and not worth the effort. Quartz has a hardness of seven. Most sand is made of quartz. Diamonds have a hardness of ten. Nothing else is that hard.

After my driveway was filled with this waste material, I stored the overflow in five gallon plastic buckets that usually store cooking grease in restaurants and fast food places. You find yourself in the most interesting places when you take up prospecting, even scrounging behind eateries. People throw away a lot of strange but useful objects. You may be able to tell, I'm cheap.

First, let me explain the best way to search for diamonds. Luckily, I had a round plastic plate used to set large flower pots on . It had edges that stuck up at right angles from the bottom, about an inch high. Always use blue lids for sorting. I cannot stress this point enough. Always use blue, even if you have to paint it with enamel. This is very important.

First take a partial handful of sediment and place it in the middle of your pan – then shake the pan to evenly distribute the stones. If you can see blue between the stones, you have done well. I inspected the stones and picked out anything that looks different. Sitting in the sunlight while you are doing this helps, since diamonds reflect light. Normally, I didn't know if I had found any diamonds, but at least I had some suspect stones. I continued to work this site until I discovered it was working me too much. I decided to stop work on this site on Clendennin Creek.

On a shorter trip, we checked out Big Stony Creek and found a place not far downstream from where North Fork Creek joins Big Stony Creek. This location was close to the car and it was easy work carrying the five gallon buckets of stones to sort them near the car. I started to find some interesting stones, so my pile grew faster than the first site. Something odd happened. I found a red stone that I thought was a teaberry that has a nice peppermint taste. But it felt hard.

We decided to buy a diamond detector. When you go to a jewelry store, they use them to determine if a stone is a diamond or a fake. Diamonds are good conductors of heat, unlike quartz, which is a poor conductor. The diamond detector consists of a battery-driven devise which has two metal points a slight distance apart. Heat is sent to one point and if it conducts heat, the other point will pick it up and a light runs up the side of the detector. This is easy to use with a flat polished surface, like a gem facet, but not so much with a raw uncut stone.

With raw diamonds, the surface is smooth but not necessarily as flat as a cut diamond facet. It takes a little time, while being very gentle, to get a slight reading that this is a diamond. In my amateur way, I might be the first diamond prospector to find diamonds with this method on raw stones.

Once I picked up a red stone and noticed that the red was peeling off, revealing an orange stone base, which revealed itself to be a diamond. This was in 1998, and I was not familiar with the diamond market, at the time. I did contact a jeweler in New Your City, and sent him 39 carats which he held for me. I found out that in the US, they will buy rough diamonds if they are worth cutting.

A word to the wise, about putting facets on diamonds. It takes a great deal of skill to cut facets on a diamond, which drives its value up a lot. Diamonds are extremely hard, but will chip off if struck just right. After cutting facets on a raw diamond, each facet is polished with diamond dust to a high shine. This can double or triple the value of your stone. This is on the wholesale market. When you set the stone in gold or silver jewelry for retail sale, it hits its highest value. Be sure to ask the retailer to check any diamonds you buy retail, with a diamond detector. You want to be sure that you are buying the real thing.

The story of the orange diamond that I found, didn't pan out as well as I hoped. I found a place in Buffalo, NY where they agreed to test my rough diamond. I called them later to see what they found, only to have my stone rudely dismissed as a piece of quartz. They had tested it with molten potassium hydroxide which will dissolve glass and quartz, but will not damage a diamond. They claim it melted, but I just knew they were lying to me. I had no money to pursue the case, so I let it go unchallenged. Few can be trusted.

Some ten years later, I was reading that in 1998, a .85 carat red diamond was found which auctioned off for a million dollars due to its rarity. I don't know if an orange diamond had ever been discovered at that time, but it could have probably brought that price or more with the craze for colored stones. Those days are gone, and the love of colored diamonds has waned, along with their values. They are still more valued than clear diamonds. If only I had the computer and resources then, I would be rich today. Technical experience is a hard thing to come by even with all my education.

My experience prospecting diamonds was getting too addictive. Since there was a good deposit of gravel at the new site, I learned a lot more about what not to do. It makes logical sense for the smaller diamonds to be washed off the mountain, while the larger ones were slower to descend. I would guess that this erosion of the mountain, with diamonds washing away in stream beds, has been going on for a short time, geologically speaking. On one trip to the top of Peters Mountain, I discovered a layer of iron ore (hematite). Apparently, the top of the TS had not been exposed all that long. A larger stone could wedge itself between other rocks and resist being carried away in high water, while the lighter material would be swept away, sinking the stones ever deeper. All will eventually be washed away in the millions of years to come.

I had a helper with me on several prospecting trips. He was my wife's eldest son, Jerry or "Topper", a very quiet man, but the best worker I ever had come with me. We decided to take the whole gravel bar apart, and many round trips were made to produce about 25 gallons of sediment to dry and sift through together. Unfortunately, we had little sparkle to show for it. Logic pointed to heading further up stream on North Creek, as long as we could still haul full buckets of stone back from our new site. It proved heavy work even though we were close to the road and the car.

Dismantling that gravel bar was hard work. The last time we dug a three foot deep trench, it filled with water. It was an extremely cold place to work, immersed in that frigid creek water. Being November in West Virginia, no wonder it was cold. Naturally, I caught hypothermia as a result of my single-minded efforts. Luckily, I made it back to the car, turned on the heat and could only speak coherently, after an hour defrosting there. I learned to never again go prospecting in November, or when it was too cold outside. Yes, lesson one: prospecting is not worth dying for.

By this time, I had given up using the hand strainers, and devised a simple sieve made of two poles and screen. The poles were about six feet long, with a covering of coarse screening on top slung between them. Another finer mesh screen caught the smaller stones underneath. We collected the bottom layer of stones in buckets to sort at home. It was easy to check this pile by running my fingers down the pile, pushing the quartz off to the side. No large diamonds found on top, but we did find a number of smaller rough diamonds in the lower layer, with this method.

We worked the site on Peters Mountain for a year or so, closing it down in the colder months. There was a pool of clear water with fine white sand on the bottom, which was from the TS. A ledge of sandstone dammed the creek. It was a very pretty site, and easy to work. The following spring, I found a mass a limbs, cuttings, mud, and fishing line all tangled together in a messy pile. Beavers may have tried to dam the stream. It was time for a new prospecting site.

Sign just before the left turn on SR 613 and decent place to camp overnight

Not too far upstream, the North Fork flows into Big Stony Creek. We found a good deposit and started to uncover more small diamonds. Some places in the cut, you could see thin layers of white sand and a greater probability of stones. This turned out to be the best site so far. It just made logical sense to me that larger stones were further up North Fork. All the work would have to be done on site, to hit possible 'pay dirt'.

Topper and I walked up North Fork on a path that used to have steel rails for timbering. The path ended and was marked by a steel wheel. A circle of rocks lay around a place to camp. Apparently, this was where hikers on the Appalachian Trail had camped before. We hid the mattock pick and some other thing for the next year. A mattock is fantastic for digging a tight stream bed.

That winter I fell ill, running a high temperature that landed me in the hospital in Princeton. My prostate trouble got so bad that I was transferred to the hospital in Bluefield, WV. Topper was seriously ill as well and joined me in the hospital. Little was done for us there; Topper died, followed soon after by my wife, who had many ailments and suffered from multiple sclerosis. Somehow, I survived and headed up to Ellamore, WV to rent a room from my old friend Maurice Griffin.

Ellamore was full of adventurous characters. One trip I teamed up with Tommy Thacker who wanted to learn prospecting from me. We went over the mountain and down to the stream that joins North Fork, and another trip to the top of Peters Mountain with him. Another stream there, joins this triple junction and flows from north to south. This was an interesting stream that had all kinds of nice deposits to work. We worked one day and found nothing worthwhile, but it was a nice place to camp between two mountains.

Tommy hiked upstream and found a beaver dam. I started to feel a lot of pain, so we left for the junction, and walked down the North Fork. I cached my mattock and pans in a safe place, and shortly after that found the camp and path out. I really needed to see a doctor soon for my prostate troubles again. As fate would have it, I fell in a beaver dam to top off my miseries. I had no injuries from the fall, but it was a very large dam to escape. My second lesson on what not to do while prospecting: misery loves company. You are damned if you leave early, and "dammed" if you don't.

The year after my move to Ellamore, a number of attempts were made to go prospecting again, but nothing came of it. Finally, I found the man I wanted to team up with, Mark Gasper, for my next diamond prospecting trip. We studied some forest maps that showed us a dirt road that wound its way up the western side of the mountain, and then down the eastern side. The lower part, on the east side, had been closed off to traffic probably to save the terrain from four-wheelers tearing up the dirt roads.

Since my old friend Lloyd Cowling lived in Narrows (not far from Peters Mountain), he said that he would drive us to the top of the mountain. We offloaded cases of water bottles and walked about a mile down from the top of Peters Mountain to a rest area for hikers. The purpose of this trip was to sift sand under a large exposed cliff. How far down diamonds could be found was strong in my mind, but the road looked pretty bad, and the weather was very hot, so we decided to cancel the whole adventure. This could be saved for cooler days.

When autumn arrived, I drove back down to Narrows in a borrowed pickup truck with a rider mower in the rear and a small trailer tossed on top of it all. Lloyd volunteered to drop me off on the mountain, so I could drive the mower to the top this time, around several gates on the forest dirt roads. I barely set up camp before other campers complained to the forest rangers about me. Evidently, I was not allowed to drive the mower up the trails and had to make a fast exit the next day. Later, I learned that the ranger would be willing to give me a ride to the top in his truck, if I set it up in advance. Live and learn. Ask and you may get a free ride.

Sometime back, I found two small diamonds under the TS on top of East River Mountain where Rt. 52 (now it's VA SR 598 and WV 598) crosses. They were small stones but hinted about how deep in the TS diamonds go. The nice thing about prospecting there is that it's all TS rock, without all the other sediments mixed with TS sand.

It seems like Global Warming is already here. In recent years, drought-like conditions cut the water flow on the streams in the summers and snowfall in the winter. These conditions really drained the groundwater reserves. Hardwood forests were dying from the droughts on the Eastern Seaboard last year, and now we are seeing extraordinary rainfalls this year. Extreme weather shifts seem to characterize Global Warming. These could seriously effect stream water levels and diamond prospecting in the future.

GEOLOGICAL HISTORY

It helps to understand what the geology was like when the TS was laid down back in the past. TS layers mark the beginning of the Silurian Period which lasted about 38 million years, over 438 million years ago. Before the Silurian we had the Ordovician Period which lasted far longer.

A lot of things happened over half way through the Ordovician Period. Life had expanded considerably from the Cambrian Period, when the first fossils were found. The land suddenly became barren with no plants or animals. A number of things happened to cause this. We had an Ice Age which was extensive and deadly. Considerable numbers of plants and animals went extinct. As a consequence, the sea levels dropped. Things were starting to happen along the East Coast of North America.

A large section of land called Baltica was approaching North America as the ocean floor subducted. When Baltica met America, the pressure formed a growing mountain chain near the East Coast and was later called the Taconic orogeny . As the range grew, water flowed westward and filled an inland sea that was extensive. Present day West Virginia, Ohio, Kentucky and some other surrounding areas were covered with a shallow fresh water sea.

A lot of mud was washed onto this sea floor which eventually became compressed into shale. It can plainly be seen by driving over old Route 52, from Bluefield to Rocky Gap, VA. As you near the top, you can view reddish shale deposits which end with TS. How much of a time gap exists between these two eras is unknown. Until better dating methods are utilized, we cannot say for sure how old it is.

The Taconic Mountains continued to grow during the Silurian Era. Geologists have estimated they were around 10,000 feet high. At some point, sand began to compress into the Tuscarora Sandstone and the Silurian Period began. Life was still scarce on land, with a minor hint of new life forms along the sea's edge.

As sediment filled this sea, Isostatic Balance takes place. The weight of the sediment forces the land to sink. Eventually, much later in time, as these layers are eroded away, the land will slowly arise. This is happening today as Canada is rebounding from the pressure of the last Ice Age and eventually the Great Lakes will drain away, as well. Lakes are only temporary, in geological time.

Baltica was not as large a landmass as Africa and South America when they attached themselves, and collided with North America. Consequently, the Taconic Mountains grew as a result, but probably caused little deformation westward.

Somewhere during this period of time, a Lamporite eruption occurred in Rockland County, VA. The remnants of this pipe have been assayed and no diamonds were found. Pyrope Garnets were present, which normally signify diamonds. Earlier reports classified this as a Kimberlite pipe until geologists found distinctive differences in the two. They are also similar and one called them Kimberlite clan rocks. As this volcanic pipe began to erode, the sediments were carried by streams to the inland sea.

The point that I am trying to make is that diamonds could be carried by different streams or rivers to different places along the inland sea. These different rivers or streams can be pinpointed by observing the TS along the western side. The fine sand in the TS will change at points, leaving a conglomerate sequence on the face wall. There is a good chance that diamonds could be found in a stream dinging the mountain at that point. Conversely, it could be barren. It would appear that Rich Creek drains some conglomerate. After all, the Punch Jones diamond find gives us a strong hint that this is true.

Rivers are also "time limited" in geological time. As erosion continues, one stream may capture another. For instance, the New River is almost as old as the Nile River, but it was changed radically by the subsequent Ice Ages. Both of these old rivers flow north, unlike every other river formed since then, which all flow south. Why? Because the Canadian Shield rose during the last Ice Age and tilted North America so it slopes to the south.

Hence, Rich Creek flows south into the New River. The New flows north between Peters Mountain and East River Mountain. Diamond-rich sediment flows along the way. East River Mountain's streams carry the same TS deposits as the Gillespie Diamond testifies. It should be pointed out that East River Mountain peters out in Tazewell County, VA to the south. Deposits become thicker, the further north you go until you reach central West Virginia's famed Seneca Rocks which are solid up-thrust TS.

A faster method to locate TS, but more superficial, would be to walk the top of East River Mtn. and Peters Mtn. and note the presence of conglomerate rock. Any stream draining the area could have diamonds. The conglomerate could be a barren section with no diamonds at all. Sometimes luck is all you need to stumble on a find.

Before the Silurian Era, Virginia was practically flat, except where the Taconic Mountains were rising. As the sea level rose, the inland sea crept further east and the TS grew. Later on, when the Taconic Mountains eroded, Baltica moved eastward and eventually became part of Europe. Another ocean formed where the Atlantic resides today. Over 200 million years later, the ocean begins to close and Africa, plus South America collides with North America. This massive weight pushed the East Coast westward. The Taconic Mountains eroded down, acting as a plow and piled up mountain ranges in Virginia. This is undoubtedly the beginnings of the Blue Ridge Mountains, a much older part of the Appalachian Mountains today.

During the rolling up of these mountains, the pressure was intense enough to cause the thrust fault where the TS slipped over the Ordovician shale. As mentioned before, this left a line of boulders, part way down Peters and East River Mountains. It is difficult to estimate how high these mountains were originally—probably something on a par with Grand Tetons today. It took eons to erode them down to the gentle rolling slopes that they have now with TS and bits of Rose Hill Iron Ore just under their top soil today.

Usually when Kimberlite or Lamporite pipes occur there are multiple pipes evident. Some erosion of these pipes is present along the West Virginia border with Virginia. They are unfortunately, devoid of diamonds. There could be other such pipes further east in Virginia, but they haven't been located yet. One possible reason for this lack of discovery could be that people were too busy searching for gold deposits and ignored diamond prospecting sites. Ignorance is our enemy, like the early Conquistadors who raided Peru learned to their sorrow. They found a large emerald and hit it with a hammer only to have it shatter. It was NOT indestructible after all, as everyone believed at the time. They totally destroyed the gemstone.

With regard on being able to tell if the samples that you find are really diamonds or not—do not buy

Diamond Prospecting

expensive diamond detectors. Not only will you save money, but without the constant recalibration needed, expensive diamond detectors are pretty unreliable. When you consider that diamonds are the hardest substance known to man, one of the simplest tests for a diamond is to lay your sample on a flat rock and try to squash it with a bigger stone, applying gentle steady pressure. Do not slam it down on top of it, or it may shatter into tiny worthless pieces. If it is a true diamond, it will not break up like normal quartz samples. Another reasonably priced tester can be found at a large department store like Wal-Mart. It costs about $10 for a basic knife sharpener. These have tiny diamond dust coatings to sharpen knives, but can also be used to test a stone for hardness. All you do is scrape your sample over the industrial diamond coated surface and see if it leaves streaks or tiny grooves on your sample. If it streaks, then your sample is not a diamond but some other mineral. You can also put a fine edge on your knife, with this tester.

While what I have described fits prospecting for diamonds on Peters Mountain and East River Mountain in West Virginia, the techniques translate to other locations pretty well, too. Even though other areas may not have a Lamporite pipe to pick away on, other secondary deposits from such rich sources are always possible. On Peters Mountain there is a lot of territory to cover, and the timing has to be right. Late spring to early Fall are good, but you have to have sufficient rainfall to keep the streams active. With climate change upon us, I have seen this become a problem since I lived in Peterstown, WV.

PROSPECTING ON THE CHEAP

The following is a rough list of what you will need for your first diamond prospecting adventure. I personally travel light with only the basics for survival. You may take as much as you can comfortably carry in and out of the forest.

1. **Machete** – a very handy tool especially to hack away greenbriers and brush
2. **Shovel** – use one with a point, not a flat blade (too many rocks)
3. **Hoe**
4. **Wire sifter** – use only steel mesh with about ¼ inch holes, 4 ft. by 2 ft.
5. **Screen door sifter** – aluminum mesh, 4 ft. by 2 ft.
6. **Roofing nails** – the wide head is good for firmly holding down your sifters
7. **Roll of furnace foil tape** – excellent for covering broken pieces of wire on the sifters so you don't cut your hands up
8. **Needle** – a longer one is best esp. with a big eye for easy threading
9. **Dental floss** – use as thread for your needle to repair holes in your sifters and keep your teeth clean
10. **Lighters** – take several, but for those who are very thorough, a water-proof lighter is sold that consists of a magnesium bar and a

striker that sends off sparks to ignite the magnesium shavings when it's wet outside
11. **Flashlight and extra batteries** – saves you when it gets dark out
12. **Light weight tarp** – 12ft by 12ft. to use as a ground cover and work surface when it rains.
13. **Tough roll of heavy twine** – very useful for securing the tarp and hoisting food bags into trees to keep them away from wild animals at night
14. **Sleeping bag**
15. **Plastic bags** – even grocery sacks will do
16. **Plastic gallon water jug -** the local stream water is safe to drink except for the North Fork which has a nasty virus from ailing beavers. Go upstream from any beavers and its OK to drink.
17. **Water purifying tablets** – even if you never use them, take them along
18. **Hammer** – claw type is good, esp. for those roofing nails
19. **Long heavy duty screwdriver** – excellent for prying rocks up
20. **Pliers** – all kinds of uses, esp. working on the wire sifters
21. **Knee pads** – important to save your knees so you can hike out
22. **Old pair of tennis shoes for wading**
23. **Toilet paper**
24. **Gloves** – protect your hands while working the sediment in the sifters

25. **Hydrogen Peroxide** – cheap First Aid for cuts, scrapes, bites, etc. It works well as a mouth wash to stop tooth aches if you swish it between teeth
26. **Small notebook and pens**
27. **Small medicine kit** – band aids, aspirin, meds, arnica gel for achy muscles
28. **Soap** – in Ziploc bag with wash cloth
29. **Black plastic contractors bag** – can sling them full of water overhead for an instant shower at the end of the day – the black bag in the sun heats up and you just poke a few holes for the shower spray
30. **Candles** – you could lose your flashlight or need a fire starter
31. **Long nails** – to secure the wooden framework for your sifting apparatus
32. **Wire cutters**
33. **Diamond-based knife sharpener** – to test your stones for hardness - if they streak when you rub it on here, it's not a diamond
34. **Five Gallon plastic buckets** – 2 per person, nested for easy carrying
35. **Extra clothes** – think in layers in case it gets cold or wet

The choice of clothing is up to you. Don't take a lot. My suggestion is to wear a watch cap which can be pulled over your head and ears. It does gets cold on the mountain at night, especially since it's over 4,000 feet high. Take a heavy sweatshirt, an extra pair of blue jeans, extra socks and underwear, and spare camp shoes so you can take off your wet sneakers at night.

Watch the weekly weather forecasts before you schedule a prospecting trip. Avoid rainy weather. These streams on the mountain can grow in violence during heavy rains. An excavated site can cease to exist and be washed away overnight. Do not get in a stream during a storm especially during a thunderstorm. Lightning strikes are no laughing matter when you are wading in water. Flash floods are becoming more common. Inches of rain can seriously ruin a primitive camping trip and wash out roads in the area. Rock and mud slides are to be avoided.

Food is something to consider. I prefer peanut butter and canned tuna, as staples. One thing you should be aware of, is that cooking meat over fires may attract unwelcome visitors such as bears. I've never seen one myself, but my first helper, Jerry said that he saw one outside Princeton, WV. That is possible, since they have a hunting radius of around 50 miles. They normally stay away from people unless they are attracted by food smells to your campsite. It is smart to bring an extra draw-string sack to sling your food up a tree over-night. Worst case, they steal it anyhow and you might have to go fishing for dinner. Be prepared for snafus. An energy bar may be all you have left to get you back to the car.

Diamond Prospecting

Once I was camping in New Mexico and the forest ranger told me that it was so quiet the past night because a mountain lion may have been circling our tents. Don't worry so much about lions here in the east, but be cautious about bragging about the diamonds you found around strangers. The most dangerous animal on this earth is our fellow man. When he is uneducated, poor, high on drugs or alcohol—you want to keep your distance. Meth is especially dangerous because the person becomes fearless with no conscience anymore, sort of like those silent mountain lions on the prowl. They are hunting for things to steal and pawn. If you mention diamonds or show off your best stones to them, be prepared to defend yourself and your rock samples. Stupid people commit stupid crimes.

If you find somebody who wants to buy some stones from you, be ready to explain why they are real diamonds. Most people will not touch rough diamonds because they haven't got a clue how to test them. They expect cut and polished stones, not something that they would toss in the driveway. Even if your stone is a diamond, it may not register on a diamond detector unless it is unusually flat or faceted. It takes time and patience to get these devices to work on irregular surfaces.

When you are camping, keep a fire going, especially if you are working in water. The simplest fire and easiest to keep going is one where a good dry log is tossed in the middle and just let it burn up slowly. Slide two more logs in on top of that one once it's caught fire. You don't want a lot of smoke, so don't try to burn green wood. Only rash greenhorns cut down live trees to burn. Don't get laughed at by this mistake.

DIRECTIONS TO THE PLACES COVERED ABOVE

There are several ways to get to Peters Mountain. If you want to get there via Interstate 40, take the Lewisburg exit onto Rt. 219 South. This will take you to Peterstown through Union, WV. There is a plaque in Peterstown that tells about the Punch Jones Diamond. The sign is on Rich Creek.

If you want to pan Rich Creek, go up the mountain but be sure to get permission from the property owner there and get some kind of agreement with him. I have a feeling that this may be a good site or even excellent for panning diamonds.

If you want to go directly to the eastern slopes where diamonds have been found on Peters Mountain, take Route 460 East from Princeton, WV. This will take you through Rich Creek, VA where Rich Creek empties into the New River. Continue east to Narrows, then Pearisburg, VA.

You will spot Rte. 635, heading north if you turn left there. This winding road goes past a lime plant, before your bear left continuing on Big Stony Creek Road which follows the creek into the Jefferson National Forest. The creek is most impressive after a heavy rain and is very popular for trout fishing now. Get off at Rt. 722 where you cross a small bridge which is where my second prospecting site was located. All that area to the left of there has been heavily prospected for diamonds. A good number of small stones were found here on Big Stony Creek.

Slightly beyond the bridge, take a right and you will end up at a gate. Beyond the gate is an old house which I think the Jefferson National Forest rangers have a place. At the gate, you cross over North Fork Creek. My site 3 is downstream from here, a couple hundred yards. I did much better at this site, than at site 2. To get a better prospecting site, you can walk upstream from there. A little rough going for a while, but you will run into an old wire fence. Cross this and an old trail exists that used to house a railway for timbering crews.

Keep going, and eventually you will run into a campsite surrounded by a ring of rocks. An old flywheel or device from timbering is at this site. You should go higher up from here. The best way is to wade the creek. It isn't that fare, and it's the easiest way to get up North Creek. At the top, two streams running north and south come together. The side stream running north contains two long beaver dams. Take the stream straight up towards Peters Mountain next. I have never explored beyond this point, but there should be good gravel deposits to 'mine'. The stream running south from there, may also be good for prospecting. You can drive to this section by going back to Route 635 and heading north a few miles, to where Route 61 crosses this creek. I have not worked this creek, but it may be profitable.

Diamond Prospecting

Any stream coming off Peters Mountain north of this area is wide open for exploration. Further north the water off Peters Mountain flows into Potts Creek which eventually flows into the James River. A very nice diamond was found where this Creek met the James River back in the 1800s.

Back to an area I know more about—there are two other streams that lie north of Clendennin Creek, but they don't look that good. This was my first prospecting site, and I may have found a few small stones there, but it wasn't until site 2 that we purchased a diamond detector. It didn't look very promising, and actually it's a short stream that flows rapidly until it hits a flat area where a lot of the water soaks into the ground.

Further north, maybe a mile or so, another stream, the Curve Creek, has not been checked for diamond prospects, and it could be good. As the Mexicans say, 'Quien Sabe' or 'who knows?'

Further north another mile or so, is Kimbalton Creek, which to my knowledge, has never been explored for diamond prospecting either. It is possibly a good bet, but very difficult to hike to. Since the Appalachian Trail covers the top of Peters Mountain, you can more easily get to it by driving up Rt. 613 and follow an old dirt road to the top. It's not too much of a walk. I've been up that and you will find some traces of iron ore on top of the sandstone called Rose Hill Iron Ore. Once you get to the top you can head south from Peters

Mountain until you get to the top of a stream leading down to North Fork Creek.

Further north of Peters Mountain, prospecting could be profitable if you head toward Seneca Rocks State Park. The layer of TS gets thicker up north, and Seneca Rocks is composed of a solid TS outcropping.

Going to East River Mountain, at Bluefield, WV, a paved road crosses and used to be called Route 52. Near the top, you can spot Ordovician red shale deposits underneath is TS. There are some overhangs that could be screened for diamonds. I found two small diamonds here, under a ledge near the road.

There are not many streams heading off on the east side of East River Mountain, but this slope should be inspected. One thing I might add, is on the west side of Peters and East River Mountains, down toward the base, are thick layers of dolomite, full of caves.

This is a neglected area, and it is a very good prospect for diamonds because in a normal stream most smaller stones get moved quicker down steam than the larger stones, geologically speaking over long periods of time. This is not true inside a cavern. Once diamonds are deposited, they can lay there for a very long time. Once water in a cave moves into the limestone, diamonds just remain in place until all the rock above the cave erodes away. Diamonds could stay in place ten times longer inside a cave than if it was situated in a stream bed exposed to the elements.

Underground work in caves is difficult because you have to get yourself and your gear down there in some very tight spots. I used to explore caves in my youth and spent a lot of time in Beacon Cave, which has several small and one larger main stream. On one side of the stream, you can hear a rumbling sound which is water flowing from high up into a pool full of gravel. It has a nice waterfall. This pool would not be difficult to work, but the main problem is carrying the equipment to the site underground. If you do decide to explore a cave site, be sure to tell someone responsible where you are prospecting. Beacon Cave, for instance, is fairly large and has been mapped out by students from Virginia Polytechnic University (known as Virginia Tech today). Back in the 1950s I went with two friends, Don Knox and Kenny Gray to discover the cave, which was followed up by the VPI (Virginia Tech) mapping efforts.

This whole limestone dolostone layer is also at the base of Peters Mountain. Some streams disappear underground there, and should be extra good for holding so much, so long. The east side of both mountains lack a limestone layer of strata to store diamonds.

Cave work demands a high degree of organization, being very careful. Injury in a cave could be a nightmare for anyone rescuing a stranded prospector. I used to use a carbide lamp underground. These were often worn by coal miners on their helmets in the old days. Times have changed. Now they all use electric lights. Cave work could be very good for recovering diamonds. You don't have to worry about the cave collapsing like the coal miners dread in their manmade tunnels. Caves are much safer than coal mines to explore.

You will have more luck prospecting if you are fascinated by geology, read maps well, and study the diamond market. The market changes all the time. In the 1990s, colored diamonds hit big. They are still worth a lot, much more than clear diamonds today. Australia has been selling top quality industrials and even black diamonds, which are also used in industry.

Just remember that before American jewelers buy your stones, they should be two or more carats in size and without flaws. Third world countries may purchase them, but I have no knowledge of how that is done. Israel might be interested because I think they have high tech laser cutters who cut and facet diamonds there. If you can't find a ready market for your rough stones, save them for future demand. Their day is coming when they will be worth far more than gold.

One other prospecting technique, that I haven't mentioned to you involves using a special black flood light. One place this would work very well is the top of the mountain where there is heavy deposits of TS. Since the top layer of TS contains diamonds, and with its downward slant, it is usually clear of vegetation—one could shine the black-light flood on the exposed rock and watch for flashes of diamonds.

I have no idea how much the very tiny diamonds will shine (not all diamonds reveal themselves on the surface, but maybe enough so this is a practical technique). If a large glow occurs, you might have found a big diamond. Very carefully pry it out of the conglomerate with an ice pick and soft blows with a hammer around the perimeter of the exposed diamond.

For a stream below it might be barren of larger diamonds and not worth exploring. If good readings (flashes), then a stream below may be profitable.

Both East River and Peters Mountains could be covered this way in a few weeks, prospecting along their ridges. GPS markings should be employed. Both mountains are essentially the same ridge cut in two by the New River. Since TS covers such a large area, it would be prudent to check all TS outcrops, even Seneca Rock. It is possible that other Lamporite extrusions could have occurred east of the TS area we have covered so far. This is highly possible since diamonds in the United States are just starting to gain some momentum. It's like 'the beginning of a diamond rush' rather than just another 'gold rush'.

TECTONIC THEORY OF DIAMOND FORMATION

This is an unorthodox theory of diamond formation. It specifically points to the source of carbon in diamonds. I link this to carbonaceous chondrites which contain around 3.5% carbon. These meteorites are very common. The majority of them fall into oceans which occupy most of the earth's surface. They fall in swarms which fall to the ocean floor in a pattern. Over millions of years, the floors of the oceans were peppered with meteorites hither and yon.

When subduction started, the ocean floor and the lithosphere, 40 or 50 miles below the surface, broke and slowly plunged into the earth. Once it sunk below the lithosphere, it entered the anesthosphere for a distance, before it struck the edge of a continent. During movement through the anesthosphere it was very hot and somewhat viscous, though under intense pressure. The anesthosphere was rubbed off on top of the ocean floor, which was much cooler. Meteorites along with the majority of carbonaceous chondrites were pasted to the top of the ocean floor.

Geologists have calculated that 90 to 120 miles deep, the temperatures and pressure are ideal for diamond formation from carbon. From 90 to 100 miles down, cubic diamonds formed, 100 to 110 miles, eight-sided ones are produced, and from

110 to 120 miles deep, twelve-sided diamonds form, roughly speaking.

When the continental masses met, just short of the 90 mile beginning of diamond formation, the first underplating of the continent began in the lower depths. These carbonaceous chondrites become industrial diamonds. The largest portion of diamonds were deposited here.

Underplating continued past the depth of 120 miles, and finally into the anesthosphere, where eventually the slab disappears. Jordan is probably correct, when he listed the roots of continents far deeper than some think or theorize.

Once subduction stopped, and the thick slab detached from the subduction zone, and descended deeper, the anesthosphere slowly took its place up against the roots of the continent.

Possibly, the hotter anesthosphere came in contact with the cooler regions of the continental root, a gap could have occurred along with the unevenness of the continent. In any event, some kind of a conduit or conduits appeared. Evidence for this existed from inclusions found in diamonds. Some inclusions could only have occurred at much deeper than 120 miles down. Large amounts of carbon dioxide and methane, intensely hot and under immense pressure, could transport bits of minerals upward. Hot liquid gasses could have been destructive to the walls of conduits and increased their size.

Other evidence for conduits could be derived from this and diamonds with eight or twelve sides, with cubic protrusions growing on it, all could have grown under these conditions. In any event, this pathway was used by Kimberlite or Lamporite extrusions.

Kimberlite or lamporite pipes probably started well below the 120 mile depth, as the pressure of gasses below it pushed the kimberlite or lamporite pipes slowly upward. With increased temperature and pressure the basalt is converted to eclogite, which is denser and shrinks during the formation process. eclogite is found in eruptions and contains diamonds. When eclogite formed, it could have resulted in a deep-seated earthquake which could have opened up conduits for the rush to the surface.

As this passed the 90 mile point where cubic diamonds form, it could have picked up the majority of industrial diamonds. Being closer to the surface faults, etc. could have suddenly released pressure, and with explosive energy, erupted.

Why are Pyrope garnets associated with diamonds? No one has answered this phenomena, but if carbonaceous chondrites are the source of carbon, the rest of the material could be responsible. Other marker minerals could have a similar origin.

Another inclusion found is a small portion of carbon dioxide. This is a rare find, but if carbon dioxide supplies the carbon for diamond growth, why is this bubble present? This may imply that methane is the carbon source. Possibly, the two gasses work together to deposit carbon. I'll put my money on methane as the carbon source as a hypothesis.

CONCLUSION

To summarize, diamonds have their origin in outer space as stony meteorites known as Carbonaceous (containing carbon) Chondrites. They fall into the oceans in a rough pattern; oceans grow for roughly 200 million years then the ocean floors eventually subduct.

Subduction first carries the slab through the anesthosphere where basalt is scraped off, covering the ocean slab. Eventually, the slab hits continental roots and is scraped off again. This is not quite deep enough, but close, to the diamond stability zone which involves proper pressure and heat to produce gem diamonds. This is an area for industrial diamond formation; not quite deep enough to reach pressures and temperatures to produce gem diamonds.

Subduction downward is assumed to produce diamonds at 90-100 miles below the surface in cubic formations. From 100-110 miles down, you find eight-sided diamonds. Below that to 120 miles deep, and 12-sided diamond formations occur. The remainder of the ocean slab eventually sinks into the anesthosphere, after leaving the continental roots.

Now, specifically to our area of interest. It was originally thought that a Kimberlite pipe erupted in Rockland County, Virginia. Later, it was classified as a Lamporite pipe and Kimberlite is reserved to shield areas where ancient cratons are cemented together by pressure. Lamporite pipes are slightly different and called Kimberlite class rocks. They both contain diamonds, hence, the earlier confusion.

Both pipes are formed deeper than the diamond stability zone where Kimberlite or Lamporite is produced and powered by massive quantities of CO_2 (Carbon Dioxide) and CH_4 (Methane) in a very hot liquid state.

I theorize that CH_4 is the source of carbon being added to coats of diamond carbon. Inclusions of minerals occur inside a diamond because liquid CO_2 was found present in a bubble, as an inclusion.

These pressurized gases push the 'lava' upward and probably help dissolve rocks to create a pathway for eruption. After picking up industrial diamonds, the 'lava' bores its way through faults and the gases expand and eventually erupt to the surface.

Going back into the Ordovician Period, a large land area called Baltica collided with the East Coast of North America. A mountain range was thrown up, called the Taconic Mountains. This corresponds with the Caledonian Mountains of Scotland and Norway in Europe.

Vast quantities of mud were deposited in the Inland Sea, producing the thick layers of shale toward the end of the Ordovician. As the Taconic Mountains rose, they lifted sand and the Tuscarora Sandstone (TS) was formed. This also marked the beginning of the Silurian Period.

The Ocean was much lower due to an extensive Ice Age that killed many plants and animals. As the ice melted, oceans started to deepen and the Inland Sea expanded eastward into present-day Virginia. As the Inland sea grew, there were water plants that produced oxygen as a waste product. It combined with dissolved iron and a layer of iron oxide which covered the TS in a layer called the Rose Hill Formation which extended into Virginia.

Rivers draining the diamond pipe are short-lived, geologically speaking, and can be determined by inspection of the TS on the west side. Rivers containing diamonds should leave conglomerate areas scattered. If a stream drains this area, diamonds could be present. On the east side of Peters and East River Mountains, diamonds eventually enter the New River or Bluestone River, much too diluted in concentration for panning economically.

The West side of the mountains is different since at their base there is a thick layer of limestone or dolomite. This produces a karst area with extensive caves. Some of the run-off from this side carries diamonds underground where they will stay until erosion eventually dissolves the karst formations. Cave exploration could reveal good deposits of diamonds. This would be dangerous work unless you are an experienced cave explorer. Virginia Tech in Blacksburg, VA is the best-equipped for this kind of prospecting. They have been spelunking for decades.

Around 250 million years ago, another subduction occurred as the continents Africa and South America collided, and the North American East Coast experienced enormous pressure and squeezed rock layers into folded mountain ranges ending with Peters and East River Mountains. The pressure was so great that the Taconic Mountains moved westward, acting as a giant plow to push up the valley and ridge portion of the state of Virginia. The "plow" is called the Blue Ridge Mountains today.

I might add that my description of sifting deposits would take a longer period to hunt for colored diamonds which would be marketable. Just the 1/4 inch screen sieve would produce sellable diamonds in the US market. Smaller ones could be slightly economical if cut in the third world countries. India is the primary place to have facets cut. Up to 800,000 Indians worked on the diamonds shipped from Australia. I was informed that it still goes on with many more people doing this job.

Virginia may have numerous deposits of diamonds yet undiscovered where the TS is exposed on the valley and ridge portions of the state. Possibly, we may have the beginnings of a "diamond rush" ahead. There are on both the mountains I discussed, about 80 miles to prospect for diamonds. Good luck!

Diamond Formation Theory

From space, Meteors ⟶ Ocean Fall, to the Sea Floor ⟶ Subduction, tectonic plate diving, one under another and downward ⟶ Diamond Formation, 90-120+ Miles ⟶ Picked up and travel to surface via Kimberlite or Lamporite Pipes ⟶ Reach Surface, Stream Erosion, rain fall, etc.

⟶ Prospecting, Yum!

APPENDIX A

CONTACT INFORMATION

Any questions that you may have about diamond prospecting, please feel free to write to me at the following address:

Dr. Ronald N. Bone
HC 71 Box 87
Ellamore, WV 26267

If you are interested in Emerald prospecting in North Carolina, I am open to consultation, also.

APPENDIX B

BIBLIOGRAPHY

Boyd, F. R., and Gurney, J. J. (1985). Mantle Xenoliths. Nature, 315, 387.

Bruton, Eric (1978). *Diamonds* 2e. Chilton Books Co., Radnor, PA.

Burke, K. Kidd, W. & Kusky, T. (1986). Archean Foreland Basin Tectonics in the Witwatersrand Basin, S. Africa. Tectonics, v.5, 439-456.

Cloos, Ernst (1989). Microtectonics along the Western Edge of the Blue Ridge, Maryland and Virginia. Johns Hopkins Press, Baltimore.

Dawson, J.B. (1980). Kimberlites and their Xenoliths. Springer-Verlag, New York.

Oreskes, Naomi (2003). Plate Tectonics: Insiders History of the Modern Theory of the Earth. Westview Press: Boulder, CO.

Engelder, Terry (1989). Structures of the Appalachian Foreland Fold-Thrust Belt. Wiley-AGU, Hoboken, NJ.

Field, J.E. (1979). *The Properties of Diamonds*. Academic Press, London.

Glover, L. (1989). *Tectonics of the Virginia Blue Ridge and Piedmont*. American Geophysical Union, Washington, D.C.

Haggerty, S.E. (1986). Diamond Genesis in the Multiply-constrained Model. Nature 320: 34-38.

Haggerty, S. E. (1999). A Diamond Trilogy: Superplumes, Supercontinents, and Supernovae. Earth and Planetary Science, 285.

Harlow, George (1998). *The Nature of Diamonds*. Cambridge Univ. Press, Cambridge, UK.

Jordan, T.H., The continental lithosphere, Rev. Geoph. Space Phys., 13, 1-12, 1975.

Mitchell, R. H. (1986). *Kimberlites: Mineralogy, Geochemistry and Petrology*. Plenum, New York.

Mitchell, R. H., and Bergman, S. C. (1991). *Petrology of Lamporites*. Plenum, New York.

Nixon, P.H. (1987). *Mantle Xenoliths*. Wiley, New York.

Orlov, Y.L. (1977). *The Mineralogy of Diamonds*. Wiley, New York.

Richardson, S. H. (1987). *Mantel Xenoliths*. Wiley, New York.

Richardson, S.H., Gurney, J.J., Erlank, A. J., and Harris, J.W. (1984). Origin of Diamonds in Old Enriched Mantle. Nature v. 310: 198-202.

Wilson, Arthur (1982). *Diamonds from Birth to Eternity*. Gemological Inst of America, Carlsbad, CA.

APPENDIX C

Glossary of Scientific Terms

Anesthosphere

Upper layer of the earth's mantle. Semi-liquid but still able to conduct seismic waves. Just below the lithosphere. Estimated to be between 50 and several hundred miles thick.

Archean

The earlier of the two divisions of Precambrian time, the name once applied to the oldest rocks. Archean Earth was very different from our modern Earth in tectonic activity, atmosphere and life forms.

Baltica

Geologists refer to the ancient continent which existed about 1.8 billion years as Baltica. Through the movements of plate tectonics, Baltica has been parts of several minor and major continents as the modern state of the world's continents developed.

Basalt

A dense, fine grained igneous rock, dark grey to black in color. Formed as basaltic lava flows cooled rapidly.

Cambrian Period

First period of the Paleozoic Era running from about 565 to 505 million years ago. This period is marked by an explosion of aquatic life forms.

Carbonaceous Chondrites

Also referred to as C Chondrites, a type of stony chondritic meteorite (containing chondrules, i.e. round grains) consisting of eight or more groups and some non-grouped meteorites.

Craton

Comes from the Greek kratos, meaning strength. It is the stable interior part of a continental plate, having remained relatively unchanged since the Precambrian Era.

Eclogite

A greenish metamorphic rock consisting mainly of garnet and pyroxene. An unusually dense rock formed at greater pressure than typical rocks of the earth's crust.

Isostatic Balance

A state of equilibrium between the earth's lithosphere and anesthosphere. An example of non equilibrium is the Himalayan mountains which are still rising.

Karst

Karst is an area of irregular limestone formed from the dissolution of soluble rocks such as limestone, dolomite and gypsum. In this area, one can find sinkholes, caves, and underground drainage systems.

Kimberlite pipes

A carrot shaped column of deep formed material pushed up by the magma flows. Typically, these are diamond bearing formations.

Lamporite pipes

Lamporite pipes are similar to kimberlite pipes except that the boiling water and volatile compounds in the flow act corrosively on the overlying rock, thereby opening a larger upper cone. These are often martini glass in shape.

Lithosphere

The lithosphere is the solid outer layer of the earth. It consists of the uppermost mantle and the crust. It is about 60 miles thick.

Meteorite

A piece of solid debris from outer space which survives impact with the earth. In space referred to as meteoroid, passing through the atmosphere as meteor, landed on earth as meteorite.

Ordovician Period

About 500 to 425 million years ago. It follows the Cambrian Period. Life expansion and diversification continued rapidly during this era.

Pyrope Garnets

Garnets are a group of minerals with similar chemical structure and the same crystalline structure. Pyrope garnets are the only one of the family to display the blood red coloration in natural stones.

Silurian Period

This is shortest period of the Paleozoic Era. It runs from about 445 to 420 million years ago. The climate remained warm and the first true plants developed during most of the Silurian.

Subduction

The geological process where at the boundary of two tectonic plates, one plate moves down and under the other. This pushes the matter on one plate very deeply under the other. Subduction is the cause of the ring of fire around the pacific ocean. In North America, the Juan de Fuca plate is subducting under the North American plate giving rise to such volcanoes as Mt. Hood and Mt. St. Helens on the West Coast.

Taconic Mountains

A range of the Appalachian Mountains along the eastern border of New York with Connecticut, Massachusetts, and Vermont.

A look at this link will help visualize the whole idea: http://geology.teacherfriendlyguide.org/index.php?option=com_content&view=article&id=60&Itemid=80

Taconic Orogeny

The Taconic mountain range was formed in the late Ordovician Period by a collision of the North American plate into a volcanic island arc. There remains (they have worn down) are today in New England.

Tuscarora Sandstone (TS)

A mapped bedrock area in Pennsylvania, Maryland, West Virginia, and Virginia. It was deposited in the lower Silurian Period. The color is white to medium grey to gray-green. TS varies in thickness from 60 feet to over 900 feet in Pennsylvania.

Underplating

Where an oceanic plate is subducting under a continental plate a series of partial melts accumulates at the base of the crust.

About The Author

Ronald Neal Bone, received his Ph.D. in Experimental Psychology from West Virginia University. He taught Psychology at several colleges including West Virginia Wesleyan College. He has had a lifelong interest in gold, diamonds and geology.

Travel, prospecting, and new adventures have entertained him for a lifetime and more. Dr. Bone is currently retired and living in the country in Randolph County, West Virginia.

www.ingramcontent.com/pod-product-compliance
Lightning Source LLC
Chambersburg PA
CBHW071802200526
45167CB00017B/1095